51 Problems in Calculating Integrals Using *U*-Substitution with Solutions

by

Richard Shedenhelm

INTRODUCTION

There are up to nine integration techniques that the calculus student may need to learn: u-substitution, integration by parts, trigonometric integrals, trigonometric substitution, quadratic polynomials, partial fractions, rationalizing substitution, numerical integration, and improper integrals. Usually these techniques are presented in this order, but there are exceptions.

In working with students, I found it extremely helpful to separate u-substitution problems into different categories. It turns out that there are four obvious types, distinguishable on the basis of how the "u-assignment" work plays out. An example of Type 1 is

$$\int (x+1)^4\, dx,$$

where the u-assignment work looks like

$$u = x + 1$$
$$du = dx.$$

An example of Type 2 is

$$\int (3x+1)^4\, dx,$$

where the u-assignment work looks like

$$u = 3x + 1$$
$$du = 3dx$$
$$\frac{du}{3} = dx.$$

An example of Type 3 is

$$\int (3x^2 + 1)^4\, xdx,$$

where the u-assignment work looks like

$$u = 3x^2 + 1$$
$$du = 6xdx$$
$$\frac{du}{6} = xdx.$$

Type 4 problems are distinctive due to an extra "backward substitution" step. An example of such a problems is

$$\int (x+3)(x-1)^4\, dx,$$

where the u-assignment work looks like

$$u = x - 1$$
$$du = dx$$
$$u + 1 = x$$
$$u + 4 = x + 3.$$

$\left.\begin{array}{c} \\ \\ \\ \end{array}\right\}$ "backward substitution"

The main body of problems in this book are divided into the four types.

An additional aid to the new student of *u*-substitution is the identification of the four-step procedure in Types 1-3 and the five-step procedure in Type 4 problems. In problem Types 1-3 the procedure is:

1. *u*-assignment
2. *u*-substitution
3. integration in terms of *u*
4. *x*-substitution

Example: $\int (x+1)^4 \, dx$

1. *u*-assignment: $u = x + 1$
$$du = dx$$

2. *u*-substitution: $\int u^4 \, du$

3. integration: $\frac{1}{5} u^5 + C$

4. *x*-substitution: $\frac{1}{5}(x+1)^5 + C$

In Type 4 problems, the procedure is

1. *u*-assignment
2. back substitution
3. *u*-substitution
4. integration in terms of *u*
5. *x*-substitution

Example: $\int (x+3)(x-1)^4 \, dx$

1. *u*-assignment: $u = x - 1$
$$du = dx$$

2. back substitution: $u = x - 1$
$$u + 4 = x + 3$$

3. *u*-substitution: $\int (u+4)u^4 \, du$

4. integration: $\int (u+4)u^4 \, du = \int (u^5 + 4u^4) \, du = \int u^5 \, du + 4 \int u^4 \, du = \frac{1}{6}u^6 + \frac{4}{5}u^5 + C$

5. *x*-substitution: $\frac{1}{6}(x-1)^6 + \frac{4}{5}(x-1)^5 + C$

iv

Every solution in this book has a "check step." This optional step verifies that the answer to the integral problem is correct by finding that the derivative of the "answer" equals the original integrand. For example, if we find that

$$\int (x+1)^4 \, dx = \tfrac{1}{5}(x+1)^5 + C,$$

we can check this by:

$$\left[\tfrac{1}{5}(x+1)^5 + C\right]' = \tfrac{5}{5}(x+1)^4 = (x+1)^4 . \checkmark$$

After the solutions, there is an appendix. This appendix includes two sets of randomly-ordered problems with answer keys following them. These problems are designed to help the student test his skill and detect which types of problems need special attention. The answers immediately following the problems in random order include the number of the original problem. The first set contains problems only from Types 1-3, whereas the second set contains problems from all four types. I broke up the sets this way since many calculus classes do not include Type 4 u-substitution problems.

Athens, Georgia, June 5, 2015.

Richard Shedenhelm

PROBLEMS

Type 1

1. $\int (x+1)^4\,dx$

2. $\int (x-50)^6\,dx$

3. $\int \sqrt{x+1}\,dx$

4. $\int \sqrt[6]{x-50}\,dx$

5. $\int \frac{1}{(x+2)^3}\,dx$

6. $\int \frac{1}{(x-21)^5}\,dx$

7. $\int \frac{1}{\sqrt{x+2}}\,dx$

8. $\int \frac{1}{\sqrt[5]{x-21}}\,dx$

9. $\int \frac{1}{x+3}\,dx$

10. $\int \frac{1}{x-3}\,dx$

11. $\int \cos(x+\pi)\,dx$

12. $\int \sin(x-5)\,dx$

13. $\int e^{x+3}\,dx$

14. $\int e^{31+x}\,dx$

Type 2

15. $\int (3x+1)^4\,dx$

16. $\int \left(\frac{1}{2}x-50\right)^6\,dx$

17. $\int \sqrt{3x+1}\,dx$

18. $\int \sqrt[6]{\frac{1}{2}x-50}\,dx$

19. $\int \frac{1}{(3x+2)^3}\,dx$

20. $\int \frac{1}{\left(\frac{3}{4}x-21\right)^5}\,dx$

21. $\int \frac{1}{\sqrt{3x+2}}\,dx$

22. $\int \frac{1}{\sqrt[5]{\frac{3}{7}x-21}}\,dx$

23. $\int \frac{1}{2x+3}\,dx$

24. $\int \frac{1}{\frac{2}{5}x-3}\,dx$

25. $\int \cos(4x)\,dx$

26. $\int \sec^2\left(\frac{1}{3}x\right)\,dx$

27. $\int e^{2x+3}\,dx$

Type 3

28. $\int (3x^2 + 1)^4 x \, dx$

29. $\int \left(\frac{1}{2}x^3 - 50\right)^6 x^2 \, dx$

30. $\int x^2 \sqrt[6]{\frac{1}{2}x^3 - 50} \, dx$

31. $\int \frac{x}{(3x^2 + 2)^3} \, dx$

32. $\int \frac{x}{\sqrt{3x^2 + 2}} \, dx$

33. $\int \frac{x^2}{\sqrt[5]{\frac{3}{7}x^3 - 21}} \, dx$

34. $\int \frac{x}{2x^2 + 3} \, dx$

35. $\int \frac{x^2}{\frac{2}{5}x^3 - 3} \, dx$

36. $\int x \cos(3x^2) \, dx$

37. $\int x^2 \sin\left(\frac{2}{3}x^3 - 5\right) dx$

38. $\int \frac{x-2}{(x^2 - 4x + 3)^3} \, dx$

39. $\int e^{x^2} x \, dx$

40. $\int (x^3 + 3x)^2 (x^2 + 1) \, dx$

41. $\int \sin(x) \cos(x) \, dx$

42. $\int \cot(x) \, dx$

Type 4

43. $\int (x + 3)(x - 1)^4 \, dx$

44. $\int x^5 \sqrt[5]{1 + x^2} \, dx$

45. $\int x\sqrt{x - 1} \, dx$

46. $\int \frac{x}{\sqrt{1 + 2x}} \, dx$

47. $\int \frac{x}{\sqrt[4]{x+2}} \, dx$

48. $\int \frac{x+4}{2x+5} \, dx$

49. $\int \frac{x^2 + 4}{x + 2} \, dx$

50. $\int (x^3 + 1)^4 x^5 \, dx$

51. $\int \frac{(3 + \ln(x))^2 (2 - \ln(x))}{x} \, dx$

ANSWERS

Type 1

1. $\frac{1}{5}(x+1)^5 + C$

2. $\frac{1}{7}(x-50)^7 + C$

3. $\frac{2}{3}(x+1)^{\frac{3}{2}} + C$

4. $\frac{6}{7}(x-50)^{\frac{7}{6}} + C$

5. $\frac{-1}{2(x+2)^2} + C$

6. $\frac{-1}{4(x-21)^4} + C$

7. $2\sqrt{x+2} + C$

8. $\frac{5}{4}(x-21)^{\frac{4}{5}} + C$

9. $\ln|x+3| + C$

10. $\ln|x-3| + C$

11. $\sin(x+\pi) + C$

12. $-\cos(x-5) + C$

13. $e^{x+3} + C$

14. $e^{31+x} + C$

Type 2

15. $\frac{1}{15}(3x+1)^5 + C$

16. $\frac{2}{7}\left(\frac{1}{2}x - 50\right)^7 + C$

17. $\frac{2}{9}\sqrt{(3x+1)^3} + C$

18. $\frac{12}{7}\sqrt[6]{\left(\frac{1}{2}x - 50\right)^7} + C$

19. $\frac{-1}{6(3x+2)^2} + C$

20. $\frac{-1}{3}\left(\frac{3}{4}x - 21\right)^{-4} + C$

21. $\frac{2}{3}\sqrt{3x+2} + C$

22. $\frac{35}{12}\sqrt[5]{\left(\frac{3}{7}x - 21\right)^4} + C$

23. $\ln\sqrt{2x+3} + C$

24. $\ln\sqrt{\left(\frac{2}{5}x - 3\right)^5} + C$

25. $\frac{1}{4}\sin(4x) + C$

26. $3\tan\left(\frac{1}{3}x\right) + C$

27. $\frac{1}{2}e^{2x+3} + C$

Type 3

28. $\frac{1}{30}(3x^2 + 1)^5 + C$

29. $\frac{2}{21}\left(\frac{1}{2}x^3 - 50\right)^7 + C$

30. $\frac{4}{7}\left(\frac{1}{2}x^3 - 50\right)^{\frac{7}{6}} + C$

31. $\frac{-1}{12}(3x^2 + 2)^{-2} + C$

32. $\frac{1}{3}\sqrt{3x^2 + 2} + C$

33. $\frac{35}{36}\left(\frac{3}{7}x^3 - 21\right)^{\frac{4}{5}} + C$

34. $\frac{1}{4}\ln(2x^2 + 3) + C$

35. $\frac{5}{6}\ln\left|\frac{2}{5}x^3 - 3\right| + C$

36. $\frac{1}{6}\sin(3x^2) + C$

37. $-\frac{1}{2}\cos\left(\frac{2}{3}x^3 - 5\right) + C$

38. $\frac{-1}{4}(x^2 - 4x + 3)^{-2} + C$

39. $\frac{1}{2}e^{x^2} + C$

40. $\frac{1}{9}(x^3 + 3x)^3 + C$

41. $\frac{1}{2}\sin^2(x) + C$

    or: $-\frac{1}{2}\cos^2(x) + C$

42. $\ln|\sin(x)| + C$

Type 4

43. $\frac{1}{6}(x - 1)^6 + \frac{4}{5}(x - 1)^5 + C$

44. $\frac{5}{32}(1 + x^2)^{\frac{16}{5}} - \frac{5}{11}(1 + x^2)^{\frac{11}{5}} + \frac{5}{12}(1 + x^2)^{\frac{6}{5}} + C$

45. $\frac{2}{5}(x - 1)^{\frac{5}{2}} + \frac{2}{3}(x - 1)^{\frac{3}{2}} + C$

46. $\frac{1}{6}(1 + 2x)^{\frac{3}{2}} - \frac{1}{2}(1 + 2x)^{\frac{1}{2}} + C$

47. $\frac{4}{7}(x + 2)^{\frac{7}{4}} - \frac{8}{3}(x + 2)^{\frac{3}{4}} + C$

48. $\frac{1}{4}(2x + 5) + \frac{3}{4}\ln|2x + 5| + C$

49. $\frac{1}{2}(x + 2)^2 - 4(x + 2) + 8\ln|x + 2| + C$

50. $\frac{1}{18}(x^3 + 1)^6 - \frac{1}{15}(x^3 + 1)^5 + C$

51. $\frac{5}{3}(3 + \ln(x))^3 - \frac{1}{4}(3 + \ln(x))^4 + C$

    or: $\frac{-25}{2}(2 - \ln(x))^2 + \frac{10}{3}(2 - \ln(x))^3 - \frac{1}{4}(2 - \ln(x))^4 + C$

10

Type 1

1. $$\int (x+1)^4\,dx \;=\; \int u^4\,du = \frac{1}{5}u^5 + C = \frac{1}{5}(x+1)^5 + C.$$

   $$\begin{cases} \text{Let } u = x+1 \\ du = dx \end{cases}$$

   Check: $\left[\frac{1}{5}(x+1)^5 + C\right]' = \frac{5}{5}(x+1)^4 = (x+1)^4.$ ✔

2. $$\int (x-50)^6\,dx \;=\; \int u^6\,du = \frac{1}{7}u^7 + C = \frac{1}{7}(x-50)^7 + C.$$

   $$\begin{cases} \text{Let } u = x-50 \\ du = dx \end{cases}$$

   Check: $\left[\frac{1}{7}(x-50)^7 + C\right]' = \frac{7}{7}(x-50)^6 = (x-50)^6.$ ✔

3. $$\int \sqrt{x+1}\,dx \;=\; \int (x+1)^{\frac{1}{2}}\,dx = \int u^{\frac{1}{2}}\,du = \frac{2}{3}u^{\frac{3}{2}} + C = \frac{2}{3}(x+1)^{\frac{3}{2}} + C.$$

   $$\begin{cases} \text{Let } u = x+1 \\ du = dx \end{cases}$$

   Check: $\left[\frac{2}{3}(x+1)^{\frac{3}{2}} + C\right]' = \frac{2}{3}\cdot\frac{3}{2}(x+1)^{\frac{1}{2}} = (x+1)^{\frac{1}{2}} = \sqrt{x+1}.$ ✔

4. $$\int \sqrt[6]{x-50}\,dx \;=\; \int (x-50)^{\frac{1}{6}}\,dx = \int u^{\frac{1}{6}}\,du = \frac{6}{7}u^{\frac{7}{6}} + C =$$

   $$\begin{cases} \text{Let } u = x-50 \\ du = dx \end{cases}$$

   $$= \frac{6}{7}(x-50)^{\frac{7}{6}} + C.$$

   Check: $\left[\frac{6}{7}(x-50)^{\frac{7}{6}} + C\right]' = \frac{6}{7}\cdot\frac{7}{6}(x-50)^{\frac{1}{6}} = (x-50)^{\frac{1}{6}} = \sqrt[6]{x-50}.$ ✔

5. $$\int \frac{1}{(x+2)^3}\,dx \;=\; \int (x+2)^{-3}\,dx = \int u^{-3}\,du = -\frac{1}{2}u^{-2} + C =$$

   $$\begin{cases} \text{Let } u = x+2 \\ du = dx \end{cases}$$

   $$= -\frac{1}{2}(x+2)^{-2} + C = \frac{-1}{2(x+2)^2} + C.$$

   Check: $\left[\frac{-1}{2(x+2)^2} + C\right]' = \left[-\frac{1}{2}(x+2)^{-2} + C\right]' = -\frac{-2}{2}(x+2)^3 =$

   $$= (x+2)^{-3} = \frac{1}{(x+2)^3}.$$ ✔

6. $\int \dfrac{1}{(x-21)^5}\,dx \;=\; \int (x-21)^{-5}\,dx \;=\; \int u^{-5}\,du \;=\; -\dfrac{1}{4}u^{-4}+C \;=\;$ $\qquad\left\{\begin{array}{l}\text{Let } u = x - 21 \\ du = dx\end{array}\right.$

$-\dfrac{1}{4}(x-21)^{-4}+C = \dfrac{-1}{4(x-21)^4}+C.$

Check: $\left[\dfrac{-1}{4(x-21)^4}+C\right]' = \left[-\dfrac{1}{4}(x-21)^{-4}+C\right]' = -\dfrac{-4}{4}(x-21)^{-5} =$

$= (x-21)^{-5} = \dfrac{1}{(x-21)^5}.\ \checkmark$

7. $\int \dfrac{1}{\sqrt{x+2}}\,dx \;=\; \int \dfrac{1}{(x+2)^{\frac{1}{2}}}\,dx \;=\; \int (x+2)^{-\frac{1}{2}}\,dx \;=\; \int u^{-\frac{1}{2}}\,du \;=\;$ $\qquad\left\{\begin{array}{l}\text{Let } u = x + 2 \\ du = dx\end{array}\right.$

$= 2u^{\frac{1}{2}}+C = 2(x+2)^{\frac{1}{2}}+C = 2\sqrt{x+2}+C.$

Check: $\left[2\sqrt{x+2}+C\right]' = \left[2(x+2)^{\frac{1}{2}}+C\right]' = \dfrac{2}{2}(x+2)^{-\frac{1}{2}} = (x+2)^{-\frac{1}{2}} =$

$= \dfrac{1}{(x+2)^{\frac{1}{2}}} = \dfrac{1}{\sqrt{x+2}}.\ \checkmark$

8. $\int \dfrac{1}{\sqrt[5]{x-21}}\,dx \;=\; \int \dfrac{1}{(x-21)^{\frac{1}{5}}}\,dx \;=\; \int (x-21)^{-\frac{1}{5}}\,dx \;=\; \int u^{-\frac{1}{5}}\,du \;=\;$ $\qquad\left\{\begin{array}{l}\text{Let } u = x - 21 \\ du = dx\end{array}\right.$

$= \dfrac{5}{4}u^{\frac{4}{5}}+C = \dfrac{5}{4}(x-21)^{\frac{4}{5}}+C.$

Check: $\left[\dfrac{5}{4}(x-21)^{\frac{4}{5}}+C\right]' = \dfrac{5}{4}\cdot\dfrac{4}{5}(x-21)^{-\frac{1}{5}} = (x-21)^{-\frac{1}{5}} = \dfrac{1}{\sqrt[5]{x-21}}.\ \checkmark$

9. $\int \dfrac{1}{x+3}\,dx \;=\; \int \dfrac{1}{u}\,du = \ln|u|+C = \ln|x+3|+C.$ $\qquad\left\{\begin{array}{l}\text{Let } u = x + 3 \\ du = dx\end{array}\right.$

Check: $[\ln|x+3|+C]' = \dfrac{1}{x+3}.\ \checkmark$

10. $\int \dfrac{1}{x-3}\,dx \;=\; \int \dfrac{1}{u}\,du = \ln|u|+C = \ln|x-3|+C.$ $\qquad\left\{\begin{array}{l}\text{Let } u = x - 3 \\ du = dx\end{array}\right.$

Check: $[\ln|x-3|+C]' = \dfrac{1}{x-3}.\ \checkmark$

11. $\int \cos(x+\pi)\,dx \;=\; \int \cos(u)\,du = \sin(u)+C = \sin(x+\pi)+C.$ $\qquad\left\{\begin{array}{l}\text{Let } u = x + \pi \\ du = dx\end{array}\right.$

Check: $[\sin(x+\pi)+C]' = \cos(x+\pi).\ \checkmark$

12. $\displaystyle\int \sin(x-5)\,dx = \int \sin(u)\,du = -\cos(u) + C = -\cos(x-5) + C.$ $\left\{\begin{array}{l}\text{Let } u = x - 5 \\ du = dx\end{array}\right.$

Check: $[-\cos(x-5) + C]' = -[-\sin(x-5)] = \sin(x-5) \, . \; \checkmark$

13. $\displaystyle\int e^{x+3}\,dx = \int e^u\,du = e^u + C = e^{x+3} + C.$ $\left\{\begin{array}{l}\text{Let } u = x + 3 \\ du = dx\end{array}\right.$

Check: $[e^{x+3} + C]' = e^{x+3} \, . \; \checkmark$

Type 2

14. $\displaystyle\int e^{31+x}\,dx = \int e^u\,du = e^u + C = e^{31+x} + C.$ $\left\{\begin{array}{l}\text{Let } u = 31 + x \\ du = dx\end{array}\right.$

Check: $[e^{31+x} + C]' = e^{31+x} \, . \; \checkmark$

15. $\displaystyle\int (3x+1)^4\,dx = \int u^4\,\frac{du}{3} = \frac{1}{3}\int u^4\,du = \frac{1}{3}\cdot\frac{1}{5}u^5 + C = \frac{1}{15}u^5 + C =$ $\left\{\begin{array}{l}\text{Let } u = 3x + 1 \\ du = 3dx \\ \frac{du}{3} = dx\end{array}\right.$

$\displaystyle = \frac{1}{15}(3x+1)^5 + C.$

Check: $\left[\frac{1}{15}(3x+1)^5 + C\right]' = \frac{5}{15}(3x+1)^4(3) = (3x+1)^4 \, . \; \checkmark$

Remark: Some books and teachers would do the u-assignment step differently as follows:

Let $u = 3x + 1$
$du = 3dx$

and then modify the original integral:

$\int(3x+1)^4\,dx = \frac{1}{3}\int(3x+1)^4 3\,dx,$

which will result in the following u-substitution:

$\frac{1}{3}\int u^4\,du.$

In over twenty years of tutoring calculus, every student with whom I discussed this alternative way of doing the problem *hates* it. I agree. I think the u-assignment technique employed in this book makes the u-substitution step as "obvious" and "mechanical" as can be.

16. $\displaystyle\int \left(\tfrac{1}{2}x - 50\right)^6\,dx = \int u^6\,2du = 2\int u^6\,du = 2\cdot\frac{1}{7}u^7 + C = \frac{2}{7}u^7 + C =$ $\left\{\begin{array}{l}\text{Let } u = \tfrac{1}{2}x - 50 \\ du = \tfrac{1}{2}dx \\ 2du = dx\end{array}\right.$

$\displaystyle = \frac{2}{7}\left(\tfrac{1}{2}x - 50\right)^7 + C.$

Check: $\left[\frac{2}{7}\left(\tfrac{1}{2}x - 50\right)^7 + C\right]' = \frac{14}{7}\left(\tfrac{1}{2}x - 50\right)^6\left(\tfrac{1}{2}\right) = \left(\tfrac{1}{2}x - 50\right)^6 \, . \; \checkmark$

13

17.

$$\int \sqrt{3x+1}\,dx = \int (3x+1)^{\frac{1}{2}}\,dx = \int u^{\frac{1}{2}}\frac{du}{3} = \frac{1}{3}\int u^{\frac{1}{2}}du = \frac{1}{3}\frac{u^{\frac{3}{2}}}{3\left(\frac{3}{2}\right)} + C = \quad \left\{ \begin{array}{c} \text{Let } u = 3x+1 \\ du = 3dx \end{array} \right.$$

$$= \frac{1}{3}\cdot\frac{2}{3}u^{\frac{3}{2}} + C = \frac{2}{9}u^{\frac{3}{2}} + C = \frac{2}{9}(3x+1)^{\frac{3}{2}} + C = \frac{2}{9}\sqrt{(3x+1)^3} + C. \qquad \frac{du}{3} = dx$$

Check: $\left[\frac{2}{9}\sqrt{(3x+1)^3} + C\right]' = \left[\frac{2}{9}(3x+1)^{\frac{3}{2}} + C\right]' = \frac{2}{9}\cdot\frac{3}{2}(3x+1)^{\frac{1}{2}}(3) =$

$$= (3x+1)^{\frac{1}{2}} = \sqrt{3x+1}. \checkmark$$

18.

$$\int \sqrt[6]{\frac{1}{2}x-50}\,dx = \int \left(\frac{1}{2}x-50\right)^{\frac{1}{6}}\,dx = \int u^{\frac{1}{6}}2du = 2\int u^{\frac{1}{6}}du = 2\frac{u^{\frac{7}{6}}}{\left(\frac{7}{6}\right)} + C = \quad \left\{ \begin{array}{c} \text{Let } u = \frac{1}{2}x - 50 \\ du = \frac{1}{2}dx \end{array} \right.$$

$$= 2\cdot\frac{6}{7}u^{\frac{7}{6}} + C = \frac{12}{7}u^{\frac{7}{6}} + C = \frac{12}{7}\left(\frac{1}{2}x-50\right)^{\frac{7}{6}} + C = \frac{12}{7}\sqrt[6]{\left(\frac{1}{2}x-50\right)^7} + C. \qquad 2du = dx$$

Check: $\left[\frac{12}{7}\sqrt[6]{\left(\frac{1}{2}x-50\right)^7} + C\right]' = \left[\frac{12}{7}\left(\frac{1}{2}x-50\right)^{\frac{7}{6}} + C\right]' =$

$$= \frac{12}{7}\cdot\frac{7}{6}\left(\frac{1}{2}x-50\right)^{\frac{1}{6}} = 2\left(\frac{1}{2}x-50\right)^{\frac{1}{6}}\left(\frac{1}{2}\right) = \left(\frac{1}{2}x-50\right)^{\frac{1}{6}} = \sqrt[6]{\frac{1}{2}x-50}. \checkmark$$

19.

$$\int \frac{1}{(3x+2)^3}\,dx = \int (3x+2)^{-3}\,dx = \int u^{-3}\frac{du}{3} = \frac{1}{3}\int u^{-3}du = \quad \left\{ \begin{array}{c} \text{Let } u = 3x + 2 \\ du = 3dx \end{array} \right.$$

$$= \frac{1}{3}\cdot\frac{-1}{2}u^{-2} + C = \frac{-1}{6}u^{-2} + C = \frac{-1}{6}(3x+2)^{-2} + C = \frac{-1}{6(3x+2)^2} + C. \qquad \frac{du}{3} = dx$$

Check: $\left[\frac{-1}{6(3x+2)^2} + C\right]' = \left[\frac{-1}{6}(3x+2)^{-2} + C\right]' = \frac{2}{6}(3x+2)^3(3) =$

$$= \frac{1}{3}(3x+2)^{-3}(3) = (3x+2)^{-3} = \frac{1}{(3x+2)^3}. \checkmark$$

20.

$$\int \frac{1}{\left(\frac{3}{4}x-21\right)^5}\,dx = \int \left(\frac{3}{4}x-21\right)^{-5}\,dx = \int u^{-5}\frac{4}{3}du = \frac{4}{3}\int u^{-5}du = \quad \left\{ \begin{array}{c} \text{Let } u = \frac{3}{4}x - 21 \\ du = \frac{3}{4}dx \end{array} \right.$$

$$= \frac{4}{3}\cdot\frac{-1}{4}u^{-4} + C = \frac{-1}{3}u^{-4} + C = \frac{-1}{3}\left(\frac{3}{4}x-21\right)^{-4} + C. \qquad \frac{4}{3}du = dx$$

Check: $\left[\frac{-1}{3}\left(\frac{3}{4}x-21\right)^{-4} + C\right]' = \frac{4}{3}\left(\frac{3}{4}x-21\right)^{-5}\frac{3}{4} = \left(\frac{3}{4}x-21\right)^{-5} =$

$$= \frac{1}{\left(\frac{3}{4}x-21\right)^5}. \checkmark$$

21. $\int \frac{1}{\sqrt{3x+2}}\,dx = \int \frac{1}{(3x+2)^{\frac{1}{2}}}\,dx = \int (3x+2)^{-\frac{1}{2}}\,dx = \int u^{-\frac{1}{2}}\frac{du}{3} =$ $\qquad$ Let $u = 3x + 2$
$du = 3dx$

$= \frac{1}{3}\int u^{-\frac{1}{2}}\,du = \frac{1}{3}\cdot\frac{2}{1}u^{\frac{1}{2}} + C = \frac{2}{3}u^{\frac{1}{2}} + C = \frac{2}{3}(3x+2)^{\frac{1}{2}} + C =$ $\qquad$ $\frac{du}{3} = dx$

$= \frac{2}{3}\sqrt{3x+2} + C.$

Check: $\left[\frac{2}{3}\sqrt{3x+2} + C\right]' = \left[\frac{2}{3}(3x+2)^{\frac{1}{2}} + C\right]' =$

$= \frac{2}{3}\cdot\frac{1}{2}(3x+2)^{-\frac{1}{2}}(3) = \frac{1}{3}(3x+2)^{-\frac{1}{2}}(3) = (3x+2)^{-\frac{1}{2}} = \frac{1}{(3x+2)^{\frac{1}{2}}} =$

$= \frac{1}{\sqrt{3x+2}}.$ ✓

22. $\int \frac{1}{\sqrt[5]{\frac{3}{7}x - 21}}\,dx = \int \frac{1}{\left(\frac{3}{7}x - 21\right)^{\frac{1}{5}}}\,dx = \int \left(\frac{3}{7}x - 21\right)^{-\frac{1}{5}}\,dx = \int u^{-\frac{1}{5}}\frac{7}{3}\,du =$ $\qquad$ Let $u = \frac{3}{7}x - 21$
$du = \frac{3}{7}dx$

$= \frac{7}{3}\int u^{-\frac{1}{5}}\,du = \frac{7}{3}\cdot\frac{5}{4}u^{\frac{4}{5}} + C = \frac{35}{12}u^{\frac{4}{5}} + C = \frac{35}{12}\left(\frac{3}{7}x - 21\right)^{\frac{4}{5}} + C =$ $\qquad$ $\frac{7}{3}\,du = dx$

$= \frac{35}{12}\sqrt[5]{\left(\frac{3}{7}x - 21\right)^{4}} + C.$

Check: $\left[\frac{35}{12}\sqrt[5]{\left(\frac{3}{7}x - 21\right)^{4}} + C\right]' = \left[\frac{35}{12}\left(\frac{3}{7}x - 21\right)^{\frac{4}{5}} + C\right]' =$

$= \frac{35}{12}\cdot\frac{4}{5}\left(\frac{3}{7}x - 21\right)^{-\frac{1}{5}}\frac{3}{7} = \left(\frac{3}{7}x - 21\right)^{-\frac{1}{5}} = \frac{1}{\left(\frac{3}{7}x - 21\right)^{\frac{1}{5}}} = \frac{1}{\sqrt[5]{\frac{3}{7}x - 21}}.$ ✓

23. $\int \frac{1}{2x+3}\,dx = \int \frac{1}{u}\frac{du}{2} = \frac{1}{2}\int \frac{1}{u}\,du = \frac{1}{2}\ln|u| + C = \frac{1}{2}\ln|2x+3| + C =$ $\qquad$ Let $u = 2x + 3$
$du = 2dx$

$= \ln(2x+3)^{\frac{1}{2}} + C = \ln\sqrt{2x+3} + C.$ $\qquad$ $\frac{du}{2} = dx$

Check: $\left[\ln\sqrt{2x+3} + C\right]' = \left[\ln(2x+3)^{\frac{1}{2}} + C\right]' =$

$= \left[\frac{1}{2}\ln|2x+3| + C\right]' = \frac{1}{2}\cdot\frac{2}{2x+3} = \frac{1}{2x+3}.$ ✓

24. $\int \frac{1}{\frac{2}{5}x-3}\,dx = \int \frac{1}{u}\cdot\frac{5}{2}\,du = \frac{5}{2}\int\frac{1}{u}\,du = \frac{5}{2}\ln|u| + C = \frac{5}{2}\ln\left|\frac{2}{5}x-3\right| + C = \begin{cases} \text{Let } u = \frac{2}{5}x - 3 \\ du = \frac{2}{5}dx \\ \frac{5}{2}du = dx \end{cases}$

$= \ln\left(\frac{2}{5}x-3\right)^{\frac{5}{2}} + C = \ln\sqrt{\left(\frac{2}{5}x-3\right)^5} + C.$

Check: $\left[\ln\sqrt{\left(\frac{2}{5}x-3\right)^5} + C\right]' = \left[\ln\left(\frac{2}{5}x-3\right)^{\frac{5}{2}} + C\right]' =$

$= \left[\frac{5}{2}\ln\left|\frac{2}{5}x-3\right| + C\right]' = \frac{5}{2}\cdot\frac{\frac{2}{5}}{\frac{2}{5}x-3} = \frac{5}{2}\cdot\frac{2}{5}\cdot\frac{1}{\frac{2}{5}x-3} = \frac{1}{\frac{2}{5}x-3}.$ ✓

25. $\int \cos(4x)\,dx = \int \cos(u)\frac{du}{4} = \frac{1}{4}\int\cos(u)\,du = \frac{1}{4}\sin(u) + C = \begin{cases} \text{Let } u = 4x \\ du = 4dx \\ \frac{du}{4} = dx \end{cases}$

$= \frac{1}{4}\sin(4x) + C.$

Check: $\left[\frac{1}{4}\sin(4x) + C\right]' = \frac{1}{4}\cos(4x)\cdot 4 = \cos(4x).$ ✓

26. $\int \sec^2\left(\frac{1}{3}x\right)dx = \int \sec^2(u)3du = 3\int\sec^2(u)du = 3\tan(u) + C = \begin{cases} \text{Let } u = \frac{1}{3}x \\ du = \frac{1}{3}dx \\ 3du = dx \end{cases}$

$= 3\tan\left(\frac{1}{3}x\right) + C.$

Check: $\left[3\tan\left(\frac{1}{3}x\right) + C\right]' = 3\sec^2\left(\frac{1}{3}x\right)\cdot\frac{1}{3} = \sec^2\left(\frac{1}{3}x\right).$ ✓

27. $\int e^{2x+3}\,dx = \int e^u\frac{du}{2} = \frac{1}{2}\int e^u du = \frac{1}{2}e^u + C = \frac{1}{2}e^{2x+3} + C.$ $\begin{cases} \text{Let } u = 2x + 3 \\ du = 2dx \end{cases}$

   Check: $\left[\frac{1}{2}e^{2x+3} + C\right]' = \frac{1}{2}e^{2x+3}(2) = e^{2x+3}.$ ✓

Type 3

28. $\int (3x^2+1)^4 x\,dx = \int u^4\frac{du}{6} = \frac{1}{6}\int u^4 du = \frac{1}{6}\cdot\frac{1}{5}u^5 + C = \frac{1}{30}u^5 + C = \begin{cases} \text{Let } u = 3x^2 + 1 \\ du = 6xdx \\ \frac{du}{6} = xdx \end{cases}$

$= \frac{1}{30}(3x^2+1)^5 + C.$

Check: $\left[\frac{1}{30}(3x^2+1)^5 + C\right]' = \frac{5}{30}(3x^2+1)^4 6x = (3x^2+1)^4.$ ✓

29. $\int \left(\frac{1}{2}x^3 - 50\right)^6 x^2\, dx = \int u^6 \frac{2}{3}\, du = \frac{2}{3}\int u^6\, du = \frac{2}{3}\cdot\frac{1}{7}u^7 + C =$

$\qquad\qquad\qquad\qquad\qquad\qquad\qquad\qquad\qquad\qquad\qquad$ $\left\{\begin{array}{l}\text{Let } u = \frac{1}{2}x^3 - 50 \\ \quad du = \frac{3}{2}x^2\,dx \\ \quad \frac{2}{3}du = x^2\,dx\end{array}\right.$

$= \frac{2}{21}u^7 + C = \frac{2}{21}\left(\frac{1}{2}x^3 - 50\right)^7 + C.$

Check: $\left[\frac{2}{21}\left(\frac{1}{2}x^3 - 50\right)^7 + C\right]' = \frac{14}{21}\left(\frac{1}{2}x^3 - 50\right)^6 \frac{3}{2}x^2 = \left(\frac{1}{2}x^3 - 50\right)^6 x^2.$ ✓

30. $\int x^2 \sqrt[6]{\frac{1}{2}x^3 - 50}\, dx = \int \left(\frac{1}{2}x^3 - 50\right)^{\frac{1}{6}} x^2\, dx = \int u^{\frac{1}{6}}\frac{2}{3}\, du = \frac{2}{3}\int u^{\frac{1}{6}}\, du =$

$\qquad\qquad\qquad\qquad\qquad\qquad\qquad\qquad\qquad\qquad\qquad$ $\left\{\begin{array}{l}\text{Let } u = \frac{1}{2}x^3 - 50 \\ \quad du = \frac{3}{2}x^2\,dx \\ \quad \frac{2}{3}du = x^2\,dx\end{array}\right.$

$= \frac{2}{3}\cdot\frac{6}{7}u^{\frac{7}{6}} + C = \frac{4}{7}u^{\frac{7}{6}} + C = \frac{4}{7}\left(\frac{1}{2}x^3 - 50\right)^{\frac{7}{6}} + C.$

Check: $\left[\frac{4}{7}\left(\frac{1}{2}x^3 - 50\right)^{\frac{7}{6}} + C\right]' = \frac{4}{7}\cdot\frac{7}{6}\left(\frac{1}{2}x^3 - 50\right)^{\frac{1}{6}}\frac{3}{2}x^2 = x^2\left(\frac{1}{2}x^3 - 50\right)^{\frac{1}{6}}.$ ✓

31. $\int \frac{x}{(3x^2 + 2)^3}\, dx = \int \frac{1}{(3x^2 + 2)^3}x\,dx = \int \frac{1}{u^3}\frac{du}{6} = \frac{1}{6}\int\frac{1}{u^3}\,du = \frac{1}{6}\int u^{-3}\,du =$

$\qquad\qquad\qquad\qquad\qquad\qquad\qquad\qquad\qquad\qquad\qquad$ $\left\{\begin{array}{l}\text{Let } u = 3x^2 + 2 \\ \quad du = 6x\,dx \\ \quad \frac{du}{6} = x\,dx\end{array}\right.$

$= \frac{1}{6}\cdot\frac{-1}{2}u^{-2} + C = \frac{-1}{12}u^{-2} + C = \frac{-1}{12}(3x^2 + 2)^{-2} + C.$

Check: $\left[\frac{-1}{12}(3x^2 + 2)^{-2} + C\right]' = \frac{1}{6}(3x^2 + 2)^{-3}6x = \frac{x}{(3x^2 + 2)^3}.$ ✓

32. $\int \frac{x}{\sqrt{3x^2 + 2}}\, dx = \int \frac{x}{(3x^2 + 2)^{\frac{1}{2}}}\, dx = \int \frac{1}{(3x^2 + 2)^{\frac{1}{2}}}x\,dx = \int \frac{1}{u^{\frac{1}{2}}}\frac{du}{6} = \frac{1}{6}\int\frac{1}{u^{\frac{1}{2}}}\,du =$

$\qquad\qquad\qquad\qquad\qquad\qquad\qquad\qquad\qquad\qquad\qquad$ $\left\{\begin{array}{l}\text{Let } u = 3x^2 + 2 \\ \quad du = 6dx \\ \quad \frac{du}{6} = dx\end{array}\right.$

$= \frac{1}{6}\int u^{-\frac{1}{2}}\,du = \frac{1}{6}\cdot\frac{2}{1}u^{\frac{1}{2}} + C = \frac{1}{3}u^{\frac{1}{2}} + C = \frac{1}{3}(3x^2 + 2)^{\frac{1}{2}} + C =$

$= \frac{1}{3}\sqrt{3x^2 + 2} + C.$

Check: $\left[\frac{1}{3}\sqrt{3x^2 + 2} + C\right]' = \left[\frac{1}{3}(3x^2 + 2)^{\frac{1}{2}} + C\right]' =$

$= \frac{1}{3}\cdot\frac{1}{2}(3x^2 + 2)^{-\frac{1}{2}}6x = \frac{1}{6}(3x^2 + 2)^{-\frac{1}{2}}6x = \frac{x}{(3x^2 + 2)^{\frac{1}{2}}} = \frac{x}{\sqrt{3x^2 + 2}}.$ ✓

33. $\int \dfrac{x^2}{\sqrt[5]{\frac{3}{7}x^3-21}}\,dx = \int \dfrac{x^2}{\left(\frac{3}{7}x^3-21\right)^{\frac{1}{5}}}\,dx = \int \dfrac{1}{\left(\frac{3}{7}x^3-21\right)^{\frac{1}{5}}}x^2 dx =$

$\qquad\qquad$ $\left\{\begin{array}{l} \text{Let } u = \frac{3}{7}x^3 - 21 \\ du = \frac{9}{7}x^2 dx \end{array}\right.$

$= \int \dfrac{1}{u^{\frac{1}{5}}}\dfrac{7}{9}\,du = \dfrac{7}{9}\int \dfrac{1}{u^{\frac{1}{5}}}\,du = \dfrac{7}{9}\int u^{-\frac{1}{5}}du = \dfrac{7}{9}\cdot\dfrac{5}{4}u^{\frac{4}{5}}+C = \dfrac{35}{36}u^{\frac{4}{5}}+C =$

$\qquad\qquad\qquad\qquad\qquad\qquad\qquad\qquad\qquad\qquad$ $\frac{7}{9}du = x^2 dx$

$= \dfrac{35}{36}\left(\dfrac{3}{7}x^3-21\right)^{\frac{4}{5}}+C.$

Check: $\left[\dfrac{35}{36}\left(\dfrac{3}{7}x^3-21\right)^{\frac{4}{5}}+C\right]' = \dfrac{35}{36}\cdot\dfrac{4}{5}\left(\dfrac{3}{7}x^3-21\right)^{-\frac{1}{5}}\dfrac{9}{7}x^2 = \dfrac{x^2}{\left(\frac{3}{7}x^3-21\right)^{\frac{1}{5}}} =$

$= \dfrac{x^2}{\sqrt[5]{\frac{3}{7}x^3-21}}\ .\ \checkmark$

---

34. $\int \dfrac{x}{2x^2+3}\,dx = \int \dfrac{1}{2x^2+3}x\,dx = \int \dfrac{1}{u}\dfrac{du}{4} = \dfrac{1}{4}\int \dfrac{1}{u}\,du = \dfrac{1}{4}\ln|u|+C =$

$\qquad\qquad$ $\left\{\begin{array}{l} \text{Let } u = 2x^2 + 3 \\ du = 4x dx \\ \frac{du}{4} = x dx \end{array}\right.$

$= \dfrac{1}{4}\ln(2x^2+3)+C.$

Remark: The absolute value bars can be replaced by parentheses since $2x^2+3$ is greater than zero for all x.

Check: $\left[\dfrac{1}{4}\ln(2x^2+3)+C\right]' = \dfrac{1}{4}\cdot\dfrac{4x}{2x^2+3} = \dfrac{x}{2x^2+3}\ .\ \checkmark$

---

35. $\int \dfrac{x^2}{\frac{2}{5}x^3-3}\,dx = \int \dfrac{1}{\frac{2}{5}x^3-3}x^2 dx = \int \dfrac{1}{u}\dfrac{5}{6}\,du = \dfrac{5}{6}\int \dfrac{1}{u}\,du = \dfrac{5}{6}\ln|u|+C =$

$\qquad\qquad$ $\left\{\begin{array}{l} \text{Let } u = \frac{2}{5}x^3 - 3 \\ du = \frac{6}{5}x^2 dx \\ \frac{5}{6}du = x^2 dx \end{array}\right.$

$= \dfrac{5}{6}\ln\left|\dfrac{2}{5}x^3-3\right|+C.$

Check: $\left[\dfrac{5}{6}\ln\left|\dfrac{2}{5}x^3-3\right|+C\right]' = \dfrac{5}{6}\cdot\dfrac{\frac{6}{5}x^2}{\frac{2}{5}x^3-3} = \dfrac{5}{6}\cdot\dfrac{6}{5}\cdot\dfrac{x^2}{\frac{2}{5}x^3-3} = \dfrac{x^2}{\frac{2}{5}x^3-3}\ .\ \checkmark$

---

36. $\int x\cos(3x^2)\,dx = \int \cos(3x^2)x\,dx = \int \cos(u)\dfrac{du}{6} = \dfrac{1}{6}\int \cos(u)\,du =$

$\qquad\qquad$ $\left\{\begin{array}{l} \text{Let } u = 3x^2 \\ du = 6x dx \\ \frac{du}{6} = x dx \end{array}\right.$

$= \dfrac{1}{6}\sin(u)+C = \dfrac{1}{6}\sin(3x^2)+C.$

Check: $\left[\dfrac{1}{6}\sin(3x^2)+C\right]' = \dfrac{1}{6}\cos(3x^2)\cdot 6x = x\cos(3x^2)\ .\ \checkmark$

37.

$$\int x^2 \sin\left(\tfrac{2}{3}x^3 - 5\right) dx = \int \sin\left(\tfrac{2}{3}x^3 - 5\right) x^2\, dx = \int \sin(u)\,\frac{du}{2} =$$

$$= \frac{1}{2}\int \sin(u)\, du = \frac{1}{2}\cdot[-\cos(u)] + C = -\frac{1}{2}\cos(u) + C =$$

$$= -\frac{1}{2}\cos\left(\tfrac{2}{3}x^3 - 5\right) + C.$$

Check: $\left[-\frac{1}{2}\cos(\tfrac{2}{3}x^3 - 5) + C\right]' = \frac{1}{2}\sin(\tfrac{2}{3}x^3 - 5)\cdot 2x^2 =$

$= x^2\sin\left(\tfrac{2}{3}x^3 - 5\right).$ ✓

$\left\{ \begin{array}{l} \text{Let } u = \tfrac{2}{3}x^3 - 5 \\ du = 2x^2\, dx \\ \frac{du}{2} = x^2\, dx \end{array}\right.$

38.

$$\int \frac{x-2}{(x^2 - 4x + 3)^3}\, dx = \int \frac{1}{(x^2 - 4x + 3)^3}(x-2)\, dx =$$

$$= \int \frac{1}{u^3}\,\frac{du}{2} = \frac{1}{2}\int \frac{1}{u^3}\, du = \frac{1}{2}\int u^{-3}\, du = \frac{1}{2}\cdot\frac{-1}{2}u^{-2} + C =$$

$$= \frac{-1}{4}u^{-2} + C = \frac{-1}{4}(x^2 - 4x + 3)^{-2} + C.$$

Check: $\left[\frac{-1}{4}(x^2 - 4x + 3)^{-2} + C\right]' = \frac{1}{2}(x^2 - 4x + 3)^{-3}(2x - 4) =$

$$= \frac{1}{2}(x^2 - 4x + 3)^{-3}2(x-2) = (x^2 - 4x + 3)^{-3}(x-2) =$$

$= \frac{x-2}{(x^2-4x+3)^3}.$ ✓

$\left\{ \begin{array}{l} \text{Let } u = x^2 - 4x + 3 \\ du = (2x - 4)dx \\ du = 2(x-2)dx \\ \frac{du}{2} = (x-2)dx \end{array}\right.$

39.

$$\int e^{x^2}x\, dx = \int e^u\,\frac{du}{2} = \frac{1}{2}\int e^u\, du = \frac{1}{2}e^u + C = \frac{1}{2}e^{x^2} + C.$$

Check: $\left[\frac{1}{2}e^{x^2} + C\right]' = \frac{1}{2}e^{x^2}2x = e^{x^2}x.$ ✓

$\left\{ \begin{array}{l} \text{Let } u = x^2 \\ du = 2x\, dx \\ \frac{du}{2} = x\, dx \end{array}\right.$

40.

$$\int (x^3 + 3x)^2(x^2 + 1)\, dx = \int u^2\,\frac{du}{3} = \frac{1}{3}\int u^2\, du = \frac{1}{3}\cdot\frac{1}{3}u^3 + C =$$

$$= \frac{1}{9}u^3 + C = \frac{1}{9}(x^3 + 3x)^3 + C.$$

Check: $\left[\frac{1}{9}(x^3 + 3x)^3 + C\right]' = \frac{3}{9}(x^3 + 3x)^2(3x^2 + 3) =$

$$= \frac{1}{3}(x^3 + 3x)^2(3x^2 + 3) = \frac{1}{3}(x^3 + 3x)^2 3(x^2 + 1)$$

$= (x^3 + 3x)^2(x^2 + 1).$ ✓

$\left\{ \begin{array}{l} \text{Let } u = x^3 + 3x \\ du = (3x^2 + 3)dx \\ du = 3(x^2 + 1)dx \\ \frac{du}{3} = (x^2 + 1)dx \end{array}\right.$

41. $\int \sin(x)\cos(x)\,dx = \int u\,du = \dfrac{1}{2}u^2 + C = \dfrac{1}{2}\sin^2(x) + C.$

$\left\{\begin{array}{l}\text{Let } u = \sin(x)\\ \quad du = \cos(x)\,dx\end{array}\right.$

Check: $\left[\dfrac{1}{2}\sin^2(x) + C\right]' = \dfrac{2}{2}\sin(x)\cos(x) = \sin(x)\cos(x).$ ✓

Alternate Solution:

$\int \sin(x)\cos(x)\,dx = \int \cos(x)\sin(x)\,dx = \int u\dfrac{du}{-1} = -\int u\,du =$

$= -\dfrac{1}{2}u^2 + C = -\dfrac{1}{2}\cos^2(x) + C.$

$\left\{\begin{array}{l}\text{Let } u = \cos(x)\\ \quad du = -\sin(x)\,dx\\ \quad \dfrac{du}{-1} = \sin(x)\,dx\end{array}\right.$

Check: $\left[-\dfrac{1}{2}\cos^2(x) + C\right]' = -\dfrac{2}{2}\cos(x)\left[-\sin(x)\right] = \sin(x)\cos(x).$ ✓

Remark: It may seem surprising that the two solutions give two different yet correct answers. Consider, however, that by the Pythagorean Identities

$$\dfrac{1}{2}\sin^2(x) + C = \dfrac{1}{2}\left[1 - \cos^2(x)\right] + C = \dfrac{1}{2} - \dfrac{1}{2}\cos^2(x) + C = -\dfrac{1}{2}\cos^2(x) + \left(C + \dfrac{1}{2}\right).$$

Hence, if we take the derivative of the expressions at the extreme ends of the latter equation, we get the same result, viz.,

$$\sin(x)\cos(x).$$

42. $\int \cot(x)\,dx = \int \dfrac{\cos(x)}{\sin(x)}\,dx = \int \dfrac{1}{\sin(x)}\cos(x)\,dx = \int \dfrac{1}{u}\,du =$

$\left\{\begin{array}{l}\text{Let } u = \sin(x)\\ \quad du = \cos(x)\,dx\end{array}\right.$

$= \ln|u| + C = \ln|\sin(x)| + C.$

Check: $[\ln|\sin(x)| + C]' = \dfrac{\cos(x)}{\sin(x)} = \cot(x).$ ✓

Remark: A similar problem is calculating the integral of the tangent function:

$\int \tan(x)\,dx = \int \dfrac{\sin(x)}{\cos(x)}\,dx = \int \dfrac{1}{\cos(x)}\sin(x)\,dx = \int \dfrac{1}{u}\dfrac{du}{-1} =$

$\left\{\begin{array}{l}\text{Let } u = \cos(x)\\ \quad du = -\sin(x)\,dx\\ \quad \dfrac{du}{-1} = \sin(x)\,dx\end{array}\right.$

$= -\int \dfrac{1}{u}\,du = -\ln|u| + C = -\ln|\cos(x)| + C = \ln\left|\cos^{-1}(x)\right| + C =$

$= \ln\left|\dfrac{1}{\cos(x)}\right| + C = \ln|\sec(x)| + C.$

Check: $[\ln|\sec(x)| + C]' = \dfrac{\sec(x)\tan(x)}{\sec(x)} = \tan(x).$ ✓

Type 4

43. $\int (x+3)(x-1)^4\,dx = \int (u+4)u^4\,du = \int (u^5 + 4u^4)\,du =$

$\left\{\begin{array}{l}\text{Let } u = x - 1\\ \quad du = dx\\ \quad u + 4 = x + 3\end{array}\right.$

$= \int u^5\,du + \int 4u^4\,du = \int u^5\,du + 4\int u^4\,du = \dfrac{1}{6}u^6 + \dfrac{4}{5}u^5 + C =$

$= \dfrac{1}{6}(x-1)^6 + \dfrac{4}{5}(x-1)^5 + C =$

Check: $\left[\dfrac{1}{6}(x-1)^6 + \dfrac{4}{5}(x-1)^5 + C\right]' = (x-1)^5 + 4(x-1)^4 =$

$= [(x-1)+4](x-1)^4 = (x+3)(x-1)^4.$ ✓

44. $\int x^5 \sqrt[5]{1+x^2}\, dx = \int x^4 \sqrt[5]{1+x^2}\, x\, dx = \int (u-1)^2 \sqrt[5]{u}\,\dfrac{du}{2} =$

$= \dfrac{1}{2}\int (u-1)^2 \sqrt[5]{u}\, du = \dfrac{1}{2}\int (u-1)^2\, u^{\frac{1}{5}}\, du =$

$= \dfrac{1}{2}\int (u^2 - 2u + 1)\, u^{\frac{1}{5}}\, du = \dfrac{1}{2}\int \left(u^{\frac{11}{5}} - 2u^{\frac{6}{5}} + u^{\frac{1}{5}} \right) du =$

$= \dfrac{1}{2}\int u^{\frac{11}{5}}\, du - \int u^{\frac{6}{5}}\, du + \dfrac{1}{2}\int u^{\frac{1}{5}}\, du =$

$= \dfrac{1}{2}\cdot\dfrac{5}{16} u^{\frac{16}{5}} - \dfrac{5}{11} u^{\frac{11}{5}} + \dfrac{1}{2}\cdot\dfrac{5}{6} u^{\frac{6}{5}} + C = \dfrac{5}{32} u^{\frac{16}{5}} - \dfrac{5}{11} u^{\frac{11}{5}} + \dfrac{5}{12} u^{\frac{6}{5}} + C =$

$= \dfrac{5}{32} (1+x^2)^{\frac{16}{5}} - \dfrac{5}{11} (1+x^2)^{\frac{11}{5}} + \dfrac{5}{12} (1+x^2)^{\frac{6}{5}} + C .$

Check: $\left[\dfrac{5}{32} (1+x^2)^{\frac{16}{5}} - \dfrac{5}{11} (1+x^2)^{\frac{11}{5}} + \dfrac{5}{12} (1+x^2)^{\frac{6}{5}} + C \right]' =$

$= \dfrac{5}{32}\cdot\dfrac{16}{5} (1+x^2)^{\frac{11}{5}} 2x - \dfrac{5}{11}\cdot\dfrac{11}{5} (1+x^2)^{\frac{6}{5}} 2x + \dfrac{5}{12}\cdot\dfrac{6}{5} (1+x^2)^{\frac{1}{5}} 2x =$

$= (1+x^2)^{\frac{11}{5}} x - 2(1+x^2)^{\frac{6}{5}} x + (1+x^2)^{\frac{1}{5}} x =$

$= \left[(1+x^2)^{\frac{10}{5}} - 2(1+x^2)^{\frac{5}{5}} + 1 \right] (1+x^2)^{\frac{1}{5}} x =$

$= [(1+x^2)^2 - 2(1+x^2) + 1](1+x^2)^{\frac{1}{5}} x =$

$= (1 + 2x^2 + x^4 - 2 - 2x^2 + 1)x \sqrt[5]{1+x^2} =$

$= x^5 \sqrt[5]{1+x^2} .$ ✔

Let $u = 1 + x^2$
$du = 2x\, dx$
$\dfrac{du}{2} = x\, dx$

$u - 1 = x^2$
$(u-1)^2 = (x^2)^2 = x^4$

45. $\int x\sqrt{x-1}\, dx = \int x(x-1)^{\frac{1}{2}}\, dx = \int (u+1)u^{\frac{1}{2}}\, du =$

$= \int \left(u^{\frac{3}{2}} + u^{\frac{1}{2}} \right) du = \int u^{\frac{3}{2}}\, du + \int u^{\frac{1}{2}}\, du = \dfrac{2}{5} u^{\frac{5}{2}} + \dfrac{2}{3} u^{\frac{3}{2}} + C =$

$= \dfrac{2}{5} (x-1)^{\frac{5}{2}} + \dfrac{2}{3} (x-1)^{\frac{3}{2}} + C.$

Check: $\left[\dfrac{2}{5} (x-1)^{\frac{5}{2}} + \dfrac{2}{3} (x-1)^{\frac{3}{2}} + C \right]' = (x-1)^{\frac{3}{2}} + (x-1)^{\frac{1}{2}} =$

$= \left[(x-1)^{\frac{2}{2}} + 1 \right] (x-1)^{\frac{1}{2}} = [(x-1) + 1](x-1)^{\frac{1}{2}} = x(x-1)^{\frac{1}{2}} =$

$= x\sqrt{x-1} .$ ✔

Let $u = x - 1$
$du = dx$
$u + 1 = x$

46.

$$\int \frac{x}{\sqrt{1+2x}}\,dx = \int \frac{x}{(1+2x)^{\frac{1}{2}}}\,dx = \int \frac{\left(\frac{u-1}{2}\right)}{u^{\frac{1}{2}}}\frac{du}{2} = \frac{1}{2}\int \frac{u-1}{2u^{\frac{1}{2}}}\,du =$$

$$= \frac{1}{4}\int \frac{u-1}{u^{\frac{1}{2}}}\,du = \frac{1}{4}\int \frac{u}{u^{\frac{1}{2}}}\,du - \frac{1}{4}\int \frac{1}{u^{\frac{1}{2}}}\,du = \frac{1}{4}\int u^{\frac{1}{2}}\,du - \frac{1}{4}\int u^{\frac{-1}{2}}\,du =$$

$$= \frac{1}{4}\cdot\frac{2}{3}u^{\frac{3}{2}} - \frac{1}{4}\cdot\frac{2}{1}u^{\frac{1}{2}} + C = \frac{1}{6}u^{\frac{3}{2}} - \frac{1}{2}u^{\frac{1}{2}} + C =$$

$$= \frac{1}{6}(1+2x)^{\frac{3}{2}} - \frac{1}{2}(1+2x)^{\frac{1}{2}} + C.$$

Check: $\left[\frac{1}{6}(1+2x)^{\frac{3}{2}} - \frac{1}{2}(1+2x)^{\frac{1}{2}} + C\right]' =$

$$= \frac{1}{4}(1+2x)^{\frac{1}{2}}\cdot 2 - \frac{1}{4}(1+2x)^{\frac{-1}{2}}\cdot 2 = \frac{1}{2}(1+2x)^{\frac{1}{2}} - \frac{1}{2}(1+2x)^{\frac{-1}{2}} =$$

$$= \frac{(1+2x)^{\frac{1}{2}}}{2} - \frac{1}{2(1+2x)^{\frac{1}{2}}} = \frac{\sqrt{1+2x}}{2} - \frac{1}{2\sqrt{1+2x}} =$$

$$= \frac{1+2x}{2\sqrt{1+2x}} - \frac{1}{2\sqrt{1+2x}} = \frac{(1+2x)-1}{2\sqrt{1+2x}} = \frac{2x}{2\sqrt{1+2x}} = \frac{x}{\sqrt{1+2x}}\ .\ \checkmark$$

$$\begin{cases} \text{Let } u = 1 + 2x \\ du = 2dx \\ \frac{du}{2} = dx \\ u - 1 = 2x \\[6pt] \frac{u-1}{2} = x \end{cases}$$

47.

$$\int \frac{x}{\sqrt[4]{x+2}}\,dx = \int \frac{1}{\sqrt[4]{x+2}}x\,dx = \int \frac{1}{\sqrt[4]{u}}(u-2)\,du =$$

$$= \int \left(\frac{u}{u^{\frac{1}{4}}} - \frac{2}{u^{\frac{1}{4}}}\right)du = \int \left(u^{\frac{3}{4}} - 2u^{\frac{-1}{4}}\right)du = \int u^{\frac{3}{4}}\,du - 2\int u^{\frac{-1}{4}}\,du =$$

$$= \frac{4}{7}u^{\frac{7}{4}} - 2\frac{4}{3}u^{\frac{3}{4}} + C = \frac{4}{7}u^{\frac{7}{4}} - \frac{8}{3}u^{\frac{3}{4}} + C = \frac{4}{7}(x+2)^{\frac{7}{4}} - \frac{8}{3}(x+2)^{\frac{3}{4}} + C\ .$$

Check: $\left[\frac{4}{7}(x+2)^{\frac{7}{4}} - \frac{8}{3}(x+2)^{\frac{3}{4}} + C\right]' = (x+2)^{\frac{3}{4}} - 2(x+2)^{\frac{-1}{4}} =$

$$= (x+2)^{\frac{3}{4}} - \frac{2}{(x+2)^{\frac{1}{4}}} = \frac{(x+2)-2}{(x+2)^{\frac{1}{4}}} = \frac{x}{\sqrt[4]{x+2}}\ .\ \checkmark$$

$$\begin{cases} \text{Let } u = x + 2 \\ du = dx \\ u - 2 = x \end{cases}$$

48. $\displaystyle\int \frac{x+4}{2x+5}\,dx = \int \frac{1}{2x+5}(x+4)\,dx = \int \frac{1}{u}\cdot\frac{u+3}{2}\,\frac{du}{2} = \frac{1}{4}\int \frac{u+3}{u}\,du =$

$\displaystyle = \frac{1}{4}\int \left(\frac{u}{u}+\frac{3}{u}\right)du = \frac{1}{4}\int 1\,du + \frac{1}{4}\int \frac{3}{u}\,du = \frac{1}{4}\int 1\,du + \frac{3}{4}\int \frac{1}{u}\,du =$

$\displaystyle = \frac{1}{4}u + \frac{3}{4}\ln|u| + C = \frac{1}{4}(2x+5) + \frac{3}{4}\ln|2x+5| + C.$

$\left\{\begin{array}{l} \text{Let } u = 2x+5 \\ \quad du = 2\,dx \\ \dfrac{du}{2} = dx \\ u - 5 = 2x \\ \dfrac{u-5}{2} = x \\ \dfrac{u-5}{2} + 4 = x+4 \\ \dfrac{u+3}{2} = x+4 \end{array}\right.$

Check: $\left[\frac{1}{4}(2x+5) + \frac{3}{4}\ln|2x+5| + C\right]' =$

$\displaystyle = \left[\frac{1}{2}x + \frac{5}{4} + \frac{3}{4}\ln|2x+5| + C\right]' = \frac{1}{2} + \frac{3}{4}\cdot\frac{2}{2x+5} = \frac{1}{2} + \frac{3}{2(2x+5)} =$

$\displaystyle = \frac{2x+5}{2(2x+5)} + \frac{3}{2(2x+5)} = \frac{2x+8}{2(2x+5)} = \frac{2(x+4)}{2(2x+5)} = \frac{x+4}{2x+5}.\ \checkmark$

Alternate Solution:

$\displaystyle\int \frac{x+4}{2x+5}\,dx = \int \frac{u}{2u-3}\,du.$ Using polynomial division,

$\begin{array}{r} \frac{1}{2} + \frac{\frac{3}{2}}{2u-3} \\ \hline 2u-3\,|\,u+0 \\ \underline{\ \ u - \frac{3}{2}} \\ \frac{3}{2} \end{array}$ Hence,

$\left\{\begin{array}{l} \text{Let } u = x+4 \\ \quad du = dx \\ u - 4 = x \\ 2u - 8 = 2x \\ 2u - 3 = 2x+5 \end{array}\right.$

$\displaystyle\int \frac{u}{2u-3} = \int \left(\frac{1}{2} + \frac{\frac{3}{2}}{2u-3}\right)du = \frac{1}{2}\int 1\,du + \frac{3}{2}\int \frac{1}{2u-3}\,du =$

$\displaystyle = \frac{1}{2}u + \frac{3}{2}\cdot\frac{1}{2}\ln|2u-3| + C = \frac{1}{2}u + \frac{3}{4}\ln|2u-3| + C =$

$\displaystyle = \frac{1}{2}(x+4) + \frac{3}{4}\ln|2(x+4)-3| + C = \frac{1}{2}(x+4) + \frac{3}{4}\ln|2x-5| + C.$

Check: $\left[\frac{1}{2}(x+4) + \frac{3}{4}\ln|2x-5| + C\right]' = \frac{1}{2} + \frac{3}{2}\cdot\frac{1}{2x+5} = \frac{2x+5+3}{2(2x+5)} =$

$\displaystyle = \frac{2x+8}{2(2x+5)} = \frac{2(x+4)}{2(2x+5)} = \frac{x+4}{2x+5}.\ \checkmark$

49.

$$\int \frac{x^2+4}{x+2}\,dx = \int \frac{1}{x+2}(x^2+4)dx = \int \frac{1}{u}(u^2-4u+8)du =$$

$$= \int \left(u-4+\frac{8}{u}\right)du = \int u\,du - \int 4\,du + \int \frac{8}{u}\,du =$$

$$= \int u\,du - 4\int 1\,du + 8\int \frac{1}{u}\,du = \frac{1}{2}u^2 - 4u + 8\ln|u| + C =$$

$$= \frac{1}{2}(x+2)^2 - 4(x+2) + 8\ln|x+2| + C\,.$$

Check: $\left[\frac{1}{2}(x+2)^2 - 4(x+2) + 8\ln|x+2| + C\right]' =$

$$= x+2-4+\frac{8}{x+2} = x-2+\frac{8}{x+2} = \frac{(x-2)(x+2)+8}{x+2} =$$

$$= \frac{x^2-4+8}{x+2} = \frac{x^2+4}{x+2}\,. \checkmark$$

Let $u = x+2$
$du = dx$
$u-2 = x$
$(u-2)^2 = x^2$
$u^2 - 4u + 4 + 4 = x^2 + 4$
$u^2 - 4u + 8 = x^2 + 4$

50.

$$\int (x^3+1)^4 x^5\,dx = \int (x^3+1)^4 x^3 x^2\,dx = \int u^4(u-1)\frac{du}{3} =$$

$$= \frac{1}{3}\int u^4(u-1)\,du = \frac{1}{3}\int (u^5-u^4)\,du = \frac{1}{3}\int u^5\,du - \frac{1}{3}\int u^4\,du =$$

$$= \frac{1}{3}\cdot\frac{1}{6}u^6 - \frac{1}{3}\cdot\frac{1}{5}u^5 + C = \frac{1}{18}u^6 - \frac{1}{15}u^5 + C =$$

$$= \frac{1}{18}(x^3+1)^6 - \frac{1}{15}(x^3+1)^5 + C.$$

Check: $\left[\frac{1}{18}(x^3+1)^6 - \frac{1}{15}(x^3+1)^5 + C\right]' =$

$$= \frac{6}{18}(x^3+1)^5 3x^2 - \frac{5}{15}(x^3+1)^4 3x^2 = x^2(x^3+1)^5 - x^2(x^3+1)^4 =$$

$$= x^2(x^3+1)^4[x^3+1-1] = x^2(x^3+1)^4 x^3 = (x^3+1)^4 x^5.\ \checkmark$$

Let $u = x^3+1$
$du = 3x^2dx$
$\frac{du}{3} = x^2dx$
$u-1 = x^3$

51. $\displaystyle\int \frac{(3+\ln(x))^2(2-\ln(x))}{x}\,dx = \int (3+\ln(x))^2(2-\ln(x))\frac{1}{x}\,dx =$

$\displaystyle\qquad\quad \left\{\begin{array}{l} \text{Let } u = 3 + \ln(x)\\ \quad du = \frac{1}{x}dx\\ u - 3 = \ln(x)\\ 3 - u = -\ln(x)\\ 5 - u = 2 - \ln(x) \end{array}\right.$

$\displaystyle = \int u^2(5-u)\,du = \int (5u^2 - u^3)\,du = \int 5u^2\,du - \int u^3\,du =$

$\displaystyle = 5\int u^2\,du - \int u^3\,du = \frac{5}{3}u^3 - \frac{1}{4}u^4 + C =$

$\displaystyle = \frac{5}{3}(3+\ln(x))^3 - \frac{1}{4}(3+\ln(x))^4 + C.$

Check: $\left[\frac{5}{3}(3+\ln(x))^3 - \frac{1}{4}(3+\ln(x))^4 + C\right]' =$

$\displaystyle = 5(3+\ln(x))^2\left(\frac{1}{x}\right) - (3+\ln(x))^3\left(\frac{1}{x}\right) =$

$\displaystyle = \frac{5(3+\ln(x))^2 - (3+\ln(x))^3}{x} = \frac{(3+\ln(x))^2[5 - (3+\ln(x))]}{x} =$

$\displaystyle = \frac{(3+\ln(x))^2(2-\ln(x))}{x} \quad \checkmark$

Alternate Solution:

$\displaystyle\int \frac{(3+\ln(x))^2(2-\ln(x))}{x}\,dx = \int (3+\ln(x))^2(2-\ln(x))\frac{1}{x}\,dx =$

$\displaystyle\qquad\quad \left\{\begin{array}{l} \text{Let } u = 2 - \ln(x)\\ \quad du = \frac{-1}{x}dx\\ \frac{du}{-1} = \frac{1}{x}dx\\ \ln(x) + u = 2\\ \ln(x) = 2 - u\\ 3 + \ln(x) = 5 - u \end{array}\right.$

$\displaystyle = \int (5-u)^2 u\,\frac{du}{-1} = -\int (25 - 10u + u^2)u\,du =$

$\displaystyle = -\int (25u - 10u^2 + u^3)\,du = \frac{-25}{2}u^2 + \frac{10}{3}u^3 - \frac{1}{4}u^4 + C =$

$\displaystyle = \frac{-25}{2}(2-\ln(x))^2 + \frac{10}{3}(2-\ln(x))^3 - \frac{1}{4}(2-\ln(x))^4 + C.$

Check: $\left[\frac{-25}{2}(2-\ln(x))^2 + \frac{10}{3}(2-\ln(x))^3 - \frac{1}{4}(2-\ln(x))^4 + C\right]' =$

$\displaystyle = -25(2-\ln(x))\left(\frac{-1}{x}\right) + 10(2-\ln(x))^2\left(\frac{-1}{x}\right) - (2-\ln(x))^3\left(\frac{-1}{x}\right) =$

$\displaystyle = \left(\frac{25}{x}\right)(2-\ln(x)) - \left(\frac{10}{x}\right)(2-\ln(x))^2 + \left(\frac{1}{x}\right)(2-\ln(x))^3 =$

$\displaystyle = \frac{1}{x}(2-\ln(x))[25 - 10(2-\ln(x)) + (2-\ln(x))^2] =$

$\displaystyle = \frac{(2-\ln(x))}{x}[25 - 20 + 10\ln(x) + 4 - 4\ln(x) + \ln^2(x)] =$

$\displaystyle = \frac{(2-\ln(x))}{x}[9 + 6\ln(x) + \ln^2(x)] = \frac{(2-\ln(x))(3+\ln(x))^2}{x} =$

$\displaystyle = \frac{(3+\ln(x))^2(2-\ln(x))}{x} \quad \checkmark$

25

TYPE 1-3 PROBLEMS IN RANDOM ORDER

Calculate the following integrals.

1. $\int \frac{x}{(3x^2+2)^3} dx$

2. $\int \left(\frac{1}{2}x - 50\right)^6 dx$

3. $\int \frac{1}{x-3} dx$

4. $\int \frac{x}{\sqrt{3x^2+2}} dx$

5. $\int e^{x+3} dx$

6. $\int \frac{1}{\sqrt[5]{\frac{3}{7}x-21}} dx$

7. $\int e^{31+x} dx$

8. $\int \frac{x^2}{\frac{2}{5}x^3 - 3} dx$

9. $\int \sqrt{x+1} dx$

10. $\int x \cos(3x^2) dx$

11. $\int \frac{1}{\left(\frac{3}{4}x - 21\right)^5} dx$

12. $\int \frac{1}{2x+3} dx$

13. $\int (x^3+3x)^2(x^2+1) dx$

14. $\int \sqrt[6]{x-50} dx$

15. $\int \frac{x-2}{(x^2-4x+3)^3} dx$

16. $\int \frac{1}{\frac{2}{5}x-3} dx$

17. $\int \cos(x+\pi) dx$

18. $\int \frac{1}{(3x+2)^3} dx$

19. $\int \frac{1}{(x+2)^3} dx$

20. $\int \cos(4x) dx$

21. $\int (x+1)^4 dx$

22. $\int \sqrt{3x+1} dx$

23. $\int \frac{1}{\sqrt{x+2}} dx$

24. $\int \sec^2\left(\frac{1}{3}x\right) dx$

25. $\int (3x^2+1)^4 x \, dx$

26. $\int \left(\frac{1}{2}x^3 - 50\right)^6 x^2 dx$

27. $\int \frac{x}{2x^2+3} dx$

28. $\int x^2 \sin\left(\frac{2}{3}x^3 - 5\right) dx$

29. $\int \cot(x) dx$

30. $\int x^2 \sqrt[6]{\frac{1}{2}x^3 - 50} \, dx$

31. $\int \sin(x-5) dx$

32. $\int \frac{1}{\sqrt{3x+2}} dx$

33. $\int (x-50)^6 dx$

34. $\int \sin(x) \cos(x) dx$

35. $\int \frac{1}{\sqrt[5]{x-21}} dx$

36. $\int \frac{1}{x+3} dx$

37. $\int e^{2x+3} dx$

38. $\int (3x+1)^4 dx$

39. $\int \frac{1}{(x-21)^5} dx$

40. $\int \frac{x^2}{\sqrt[5]{\frac{3}{7}x^3 - 21}} dx$

41. $\int e^{x^2} x \, dx$

42. $\int \sqrt[6]{\frac{1}{2}x-50} \, dx$

ANSWERS

1. $\frac{-1}{12}(3x^2 + 2)^{-2} + C$ (#31)

2. $\frac{2}{7}\left(\frac{1}{2}x - 50\right)^7 + C$ (#16)

3. $\ln|x - 3| + C$ (#10)

4. $\frac{1}{3}\sqrt{3x^2 + 2} + C$ (#32)

5. $e^{x+3} + C$ (#13)

6. $\frac{35}{12}\sqrt[5]{\left(\frac{3}{7}x - 21\right)^4} + C$ (#22)

7. $e^{31+x} + C$ (#14)

8. $\frac{5}{6}\ln\left|\frac{2}{5}x^3 - 3\right| + C$ (#35)

9. $\frac{2}{3}(x + 1)^{\frac{3}{2}} + C$ (#3)

10. $\frac{1}{6}\sin(3x^2) + C$ (#36)

11. $\frac{-1}{3}\left(\frac{3}{4}x - 21\right)^{-4} + C$ (#20)

12. $\ln\sqrt{2x + 3} + C$ (#23)

13. $\frac{1}{9}(x^3 + 3x)^3 + C$ (#40)

14. $\frac{6}{7}(x - 50)^{\frac{7}{6}} + C$ (#4)

15. $\frac{-1}{4}(x^2 - 4x + 3)^{-2} + C$ (#3)

16. $\ln\sqrt{\left(\frac{2}{5}x - 3\right)^5} + C$ (#24)

17. $\sin(x + \pi) + C$ (#11)

18. $\frac{-1}{6(3x+2)^2} + C$ (#19)

19. $\frac{-1}{2(x+2)^2} + C$ (#5)

20. $\frac{1}{4}\sin(4x) + C$ (#25)

21. $\frac{1}{5}(x + 1)^5 + C$ (#1)

22. $\frac{2}{9}\sqrt{(3x + 1)^3} + C$ (#17)

23. $2\sqrt{x + 2} + C$ (#7)

24. $3\tan\left(\frac{1}{3}x\right) + C$ (#26)

25. $\frac{1}{30}(3x^2 + 1)^5 + C$ (#28)

26. $\frac{2}{21}\left(\frac{1}{2}x^3 - 50\right)^7 + C$ (#29)

27. $\frac{1}{4}\ln(2x^2 + 3) + C$ (#34)

28. $-\frac{1}{2}\cos\left(\frac{2}{3}x^3 - 5\right) + C$ (#37)

29. $\ln|\sin(x)| + C$ (#42)

30. $\frac{4}{7}\left(\frac{1}{2}x^3 - 50\right)^{\frac{7}{6}} + C$ (#30)

31. $-\cos(x - 5) + C$ (#12)

32. $\frac{2}{3}\sqrt{3x + 2} + C$ (#21)

33. $\frac{1}{7}(x - 50)^7 + C$ (#2)

34. $\frac{1}{2}\sin^2(x) + C$ (#41)

35. $\frac{5}{4}(x - 21)^{\frac{4}{5}} + C$ (#8)

36. $\ln|x + 3| + C$ (#9)

37. $\frac{1}{2}e^{2x+3} + C$ (#27)

38. $\frac{1}{15}(3x + 1)^5 + C$ (#15)

39. $\frac{-1}{4(x-21)^4} + C$ (#6)

40. $\frac{35}{36}\left(\frac{3}{7}x^3 - 21\right)^{\frac{4}{5}} + C$ (#33)

41. $\frac{1}{2}e^{x^2} + C$ (#39)

42. $\frac{12}{7}\sqrt[6]{\left(\frac{1}{2}x - 50\right)^7} + C$ (#18)

TYPE 1-4 PROBLEMS IN RANDOM ORDER

Calculate the following integrals.

1. $\int (3x+1)^4 \, dx$

2. $\int \frac{1}{(x+2)^3} \, dx$

3. $\int \frac{1}{\left(\frac{3}{4}x - 21\right)^5} \, dx$

4. $\int \sqrt{x+1} \, dx$

5. $\int \frac{1}{(3x+2)^3} \, dx$

6. $\int (x+1)^4 \, dx$

7. $\int \frac{1}{2x+3} \, dx$

8. $\int \frac{x^2}{\sqrt[5]{\frac{3}{7}x^3 - 21}} \, dx$

9. $\int \cos(4x) \, dx$

10. $\int \frac{x}{\sqrt{3x^2 + 2}} \, dx$

11. $\int \cot(x) \, dx$

12. $\int \frac{(3+\ln(x))^2 (2 - \ln(x))}{x} \, dx$

13. $\int \frac{x^2}{\frac{2}{5}x^3 - 3} \, dx$

14. $\int (x+3)(x-1)^4 \, dx$

15. $\int \left(\frac{1}{2}x - 50\right)^6 \, dx$

16. $\int \frac{1}{\sqrt[5]{x-21}} \, dx$

17. $\int e^{31+x} \, dx$

18. $\int \frac{1}{(x-21)^5} \, dx$

19. $\int \frac{1}{x+3} \, dx$

20. $\int \frac{1}{\sqrt{3x+2}} \, dx$

21. $\int x^2 \sqrt[6]{\frac{1}{2}x^3 - 50} \, dx$

22. $\int (3x^2 + 1)^4 x \, dx$

23. $\int \sin(x)\cos(x) \, dx$

24. $\int \frac{x+4}{2x+5} \, dx$

25. $\int (x^3 + 1)^4 x^5 \, dx$

26. $\int x \cos(3x^2) \, dx$

27. $\int \frac{x}{\sqrt{1+2x}} \, dx$

28. $\int \cos(x + \pi) \, dx$

29. $\int \sin(x - 5) \, dx$

30. $\int \frac{1}{x-3} \, dx$

31. $\int e^{x+3} \, dx$

32. $\int \frac{1}{\sqrt{x+2}} \, dx$

33. $\int \frac{1}{\sqrt[5]{\frac{3}{7}x - 21}} \, dx$

34. $\int \left(\frac{1}{2}x^3 - 50\right)^6 x^2 \, dx$

35. $\int \frac{1}{\frac{2}{5}x - 3} \, dx$

36. $\int \frac{x}{2x^2 + 3} \, dx$

37. $\int \sec^2 \left(\frac{1}{3}x\right) dx$

38. $\int \frac{x^2 + 4}{x+2} \, dx$

39. $\int \frac{(3+\ln(x))^2 (2 - \ln(x))}{x} \, dx$

40. $\int \frac{x-2}{(x^2 - 4x + 3)^3} \, dx$

41. $\int \frac{x}{\sqrt[4]{x+2}} \, dx$

42. $\int e^{x^2} x \, dx$

43. $\int \sqrt[6]{x - 50} \, dx$

44. $\int \sqrt[6]{\frac{1}{2}x - 50} \, dx$

45. $\int \sqrt{3x+1} \, dx$

46. 2. $\int (x - 50)^6 \, dx$

47. $\int \frac{x}{(3x^2 + 2)^3} \, dx$

48. $\int e^{2x+3} \, dx$

49. $\int x^5 \sqrt[5]{1 + x^2} \, dx$

50. $\int (x^3 + 3x)^2 (x^2 + 1) \, dx$

51. $\int x\sqrt{x-1} \, dx$

1. $\frac{1}{15}(3x+1)^5 + C$ (#15)

2. $\frac{-1}{2(x+2)^2} + +C$ (#5)

3. $\frac{-1}{3}\left(\frac{3}{4}x - 21\right)^{-4} + C$ (#20)

4. $\frac{2}{3}(x+1)^{\frac{3}{2}} + C$ (#3)

5. $\frac{-1}{6(3x+2)^2} + C$ (#19)

6. $\frac{1}{5}(x+1)^5 + C$ (#1)

7. $\ln\sqrt{2x+3} + C$ (#23)

8. $\frac{35}{36}\left(\frac{3}{7}x^3 - 21\right)^{\frac{4}{5}} + C$ (#33)

9. $\frac{1}{4}\sin(4x) + C$ (#25)

10. $\frac{1}{3}\sqrt{3x^2+2} + C$ (#32)

11. $\ln|\sin(x)| + C$ (#42)

12. $\frac{5}{3}(3+\ln(x))^3 - \frac{1}{4}(3+\ln(x))^4 + C$ (#

13. $\frac{5}{6}\ln\left|\frac{2}{5}x^3 - 3\right| + C$ (#35)

14. $\frac{1}{6}(x-1)^6 + \frac{4}{5}(x-1)^5 + C$ (#43)

15. $\frac{2}{7}\left(\frac{1}{2}x - 50\right)^7 + C$ (#16)

16. $\frac{5}{4}(x-21)^{\frac{4}{5}} + C$ (#8)

17. $e^{31+x} + C$ (#14)

18. $\frac{-1}{4(x-21)^4} + C$ (#6)

19. $\ln|x+3| + C$ (#9)

20. $\frac{2}{3}\sqrt{3x+2} + C$ (#21)

21. $\frac{4}{7}\left(\frac{1}{2}x^3 - 50\right)^{\frac{7}{6}} + C$ (#30)

22. $\frac{1}{30}(3x^2+1)^5 + C$ (#28)

23. $\frac{1}{2}\sin^2(x) + C$ (#41)

24. $\frac{1}{4}(2x+5) + \frac{3}{4}\ln|2x+5| + C$ (#4

25. $\frac{1}{18}(x^3+1)^6 - \frac{1}{15}(x^3+1)^5 + C$ (#50)

26. $\frac{1}{6}\sin(3x^2) + C$ (#36)

27. $\frac{1}{6}(1+2x)^{\frac{3}{2}} - \frac{1}{2}(1+2x)^{\frac{1}{2}} + C$ (#4

28. $\sin(x+\pi) + C$ (#11)

29. $-\cos(x-5) + C$ (#12)

30. $\ln|x-3| + C$ (#10)

31. $e^{x+3} + C$ (#13)

32. $2\sqrt{x+2} + C$ (#7)

33. $\frac{35}{12}\sqrt[5]{\left(\frac{3}{7}x - 21\right)^4} + C$ (#22)

34. $\frac{2}{21}\left(\frac{1}{2}x^3 - 50\right)^7 + C$ (#29)

35. $\ln\sqrt{\left(\frac{2}{5}x - 3\right)^5} + C$ (#24)

36. $\frac{1}{4}\ln(2x^2+3) + C$ (#34)

37. $3\tan\left(\frac{1}{3}x\right) + C$ (#26)

38. $\frac{1}{2}(x+2)^2 - 4(x+2) + 8\ln|x+2| + C$ (#49)

39. $\frac{5}{3}(3+\ln(x))^3 - \frac{1}{4}(3+\ln(x))^4 + C$ (#

40. $\frac{-1}{4}(x^2-4x+3)^{-2} + C$ (#38)

41. $\frac{4}{7}(x+2)^{\frac{7}{4}} - \frac{8}{3}(x+2)^{\frac{3}{4}} + C$ (#47)

42. $\frac{1}{2}e^{x^2} + C$ (#39)

43. $\frac{6}{7}(x-50)^{\frac{7}{6}} + C$ (#4)

44. $\frac{12}{7}\sqrt[6]{\left(\frac{1}{2}x - 50\right)^7} + C$ (#18)

45. $\frac{2}{9}\sqrt{(3x+1)^3} + C$ (#17)

46. $\frac{1}{7}(x-50)^7 + C$ (#2)

47. $\frac{-1}{12}(3x^2+2)^{-2} + C$ (#31)

48. $\frac{1}{2}e^{2x+3} + C$ (#27)

49. $\frac{5}{32}(1+x^2)^{\frac{16}{5}} - \frac{5}{11}(1+x^2)^{\frac{11}{5}} + \frac{5}{12}(1+x^2)^{\frac{6}{5}} + C$ (#44)

50. $\frac{1}{9}(x^3+3x)^3 + C$ (#40)

51. $\frac{2}{5}(x-1)^{\frac{5}{2}} + \frac{2}{3}(x-1)^{\frac{3}{2}} + C$ (#

www.ingramcontent.com/pod-product-compliance
Lightning Source LLC
Chambersburg PA
CBHW082305200526
45168CB00018B/3410